$15 Million Sight and Sound Theater Fire and Building Collapse
Lancaster County, Pennsylvania

Investigated by: Stanley L. Poole
Hollis Stambaugh

This is Report 097 of the Major Fires Investigation Project conducted by Varley-Campbell and Associates, Inc./TriData Corporation under contract EMW-94-C-4423 to the United States Fire Administration, Federal Emergency Management Agency.

Department of Homeland Security
United States Fire Administration
National Fire Data Center

U.S. Fire Administration Fire Investigations Program

The U.S. Fire Administration develops reports on selected major fires throughout the country. The fires usually involve multiple deaths or a large loss of property. But the primary criterion for deciding to do a report is whether it will result in significant "lessons learned." In some cases these lessons bring to light new knowledge about fire--the effect of building construction or contents, human behavior in fire, etc. In other cases, the lessons are not new but are serious enough to highlight once again, with yet another fire tragedy report. In some cases, special reports are developed to discuss events, drills, or new technologies which are of interest to the fire service.

The reports are sent to fire magazines and are distributed at National and Regional fire meetings. The International Association of Fire Chiefs assists the USFA in disseminating the findings throughout the fire service. On a continuing basis the reports are available on request from the USFA; announcements of their availability are published widely in fire journals and newsletters.

This body of work provides detailed information on the nature of the fire problem for policymakers who must decide on allocations of resources between fire and other pressing problems, and within the fire service to improve codes and code enforcement, training, public fire education, building technology, and other related areas.

The Fire Administration, which has no regulatory authority, sends an experienced fire investigator into a community after a major incident only after having conferred with the local fire authorities to insure that the assistance and presence of the USFA would be supportive and would in no way interfere with any review of the incident they are themselves conducting. The intent is not to arrive during the event or even immediately after, but rather after the dust settles, so that a complete and objective review of all the important aspects of the incident can be made. Local authorities review the USFA's report while it is in draft. The USFA investigator or team is available to local authorities should they wish to request technical assistance for their own investigation.

For additional copies of this report write to the U.S. Fire Administration, 16825 South Seton Avenue, Emmitsburg, Maryland 21727. The report is available on the Administration's Web site at http://www.usfa.dhs.gov/

U.S. Fire Administration
Mission Statement

As an entity of the Department of Homeland Security, the mission of the USFA is to reduce life and economic losses due to fire and related emergencies, through leadership, advocacy, coordination, and support. We serve the Nation independently, in coordination with other Federal agencies, and in partnership with fire protection and emergency service communities. With a commitment to excellence, we provide public education, training, technology, and data initiatives.

TABLE OF CONTENTS

OVERVIEW 1
SUMMARY OF KEY ISSUES 2
FIRE SERVICE ORGANIZATION 3
FIRE BUILDING 4
    Exit and Life Safety Features 7
    Fire Protection Features 7
FIRE ORIGIN 8
FIRE SPREAD 9
RESPONSE 9
    Fire Response 10
    Fireground Organization 10
    Defensive Operations 11
ANALYSIS 20
LESSONS LEARNED 22
BIBLIOGRAPHY 26
APPENDIX A: Lancaster CountyPennsylvania Incident Command SystemFire/ EmergencyStandard Operating Procedures 27
APPENDIX B: State Panic And Fire Code 38
APPENDIX C: Emergency Services Operating Costs 45
APPENDIX D: Lancaster Fire Chief Association Standard Operating Guidelines 47

$15 Million Sight and Sound Theater Fire and Building Collapse
300 Hartman's Bridge Road
Strasburg Township
Lancaster County, Pennsylvania 17579
January 28, 1997

Local Contacts: John D. Leas
Lancaster County Firemen's
Association Fire Service Coordinator
630 East Oregon Rd.
Lititz, PA 17543
717-560-6530 Ext. 222

William H. McCune
Pennsylvania State Trooper/Fire Marshal
2099 Lincoln Hwy. East
Lancaster, PA 17602
717-299-7663

Rick Wentz
Volunteer Fire Chief
Strasburg Township Fire Department
Mark LeFever
Volunteer Assistant Chief
Strasburg Township Fire Department

OVERVIEW

On the morning of January 28, 1997, in the Lancaster County, Pennsylvania, township of Strasburg, a fire caused the collapse of the state-of-the-art, seven-year-old Sight and Sound Theater and resulted in structural damage to most of the connecting buildings. The theater was a total loss, valued at over $15 million.

The stage area was undergoing renovation and the theater was closed to the public, however, approximately 200 people, construction staff, and employees were in the building at the time the fire started. Although the theater was built to conform to a two hour fire rated assembly code requirement, many other fire protection features that could have assisted in saving the structure and reducing the damage were not present. Further contributing to the resultant loss was the failure of the alarm system to notify the fire dispatch communications center and the lack of an adequate, readily available water supply. The volunteer fire departments that responded were faced with difficult fire conditions and tactical challenges for which they had not been adequately trained, and were without the benefit of adequate pre-planning. Local fire service suggestions for built-in fire suppression and smoke ventilation systems during the pre-construction plan review phase were ignored.

While the incident was influenced by many conditions and situations which contributed to the large fire loss, fortunately there was no loss of life and only six minor injuries to the construction company staff. If the 1,400 seat capacity auditorium had been full, the situation could have been catastrophic.

The lessons learned as a result of this fire and collapse are similar to observations from comparable incidents in recent history. The Wolftrap Farm Theater and Pavilion fire in Fairfax County, Virginia, in 1988 suffered a total loss in the stage, props, dressing rooms, and storage area. The pre-construction recommendation for a fire sprinkler system had not been heeded. When the facility was rebuilt, it was totally sprinkled.

The McCormick Place exhibition hall fire in Chicago, Illinois, in 1967 was a public assembly occupancy built with fire protected steel construction and no sprinkler system. "Fortunately the fire started in the early morning hours; the possibility of life loss would have been staggering had the fire occurred during the day."[1] This fast-burning, high-rate-of-heat-production fire caused complete collapse of the building. The fire was discovered early but the alarm was delayed while maintenance personnel tried to control the fire. The new (rebuilt) McCormick place has a hydraulically calculated sprinkler system, smoke venting system, in addition to one hour rated fire resistance protection on structural steel. The lessons learned here are not new. These examples are similar in situation and outcome to the Sight and Sound theater fire.

SUMMARY OF KEY ISSUES

Issues	Comments
Fire Origin	Welding operations caused the fire.
Sprinkler System Waived	The requirement for sprinklers in the high hazard storage area under the stage was offset for a central alarmed smoke detector system and 2-hour fire walls.
Alarm System Failure	The alarm system failed to notify the county fire communications center and contributed to a delayed response.
Employees' Delayed Report of Fire	The fire was discovered by theater employees who used fire extinguishers for several minutes before calling 9-1-1.
Structural Failure	Construction on the stage floor damaged the sprayed-on fire-resistant coating of steel structural members. The rapid fire spread caused early structural failure of the stage floor and contributed to fire extension.
Water Supply Eliminated	The water supply pond originally intended to provide water for fire suppression was eliminated to accommodate the addition of a prop manufacturing building.

---

[1] Branigan, Building Construction for the Fire Service, pg. 254

Issues	Comments
Lack of Compartmentalization	Several additions to the original theater connected the main structure to the maintenance buildings. The additional construction increased the overall size of the complex and compromised the compartmentalization. The high hazard storage area was not discretely separated from the rest of the structure.
Lack for Exterior Fire Stream Access	Lack of windows prevented exterior fire streams from being effective until the roof collapsed.
Inadequate Staff Training	The theater staff and the fire department had not trained together on managing a fire emergency in this technically and tactically challenging complex facility.
Pre-Fire Planning	The fire department was not familiar with the new additions and the increased potential risk to firefighter safety.
Fire Department Tactical Operations	The initial fire attack was conducted with under-sized handlines, inadequate for the heavy fuel load in the building.
Lack of Local Fire Code	There was no mandate for local government, fire officials, and local building owners to coordinate to ensure fire safe occupancies.

FIRE SERVICE ORGANIZATION

Strasburg, Pennsylvania, is an incorporated town located about ten miles southeast of the City of Lancaster. The population of Strasburg is approximately 10,000. The Strasburg Volunteer Fire Department services the town, the surrounding township, and the unincorporated areas outside of Strasburg. The Strasburg Volunteer Fire Department is a member of the Lancaster County Firemen's Association, which provides fire and rescue protection throughout Lancaster County. The Firemen's Association is the agency that coordinates the response and mutual aid of over 100 individual corporations, also called fire companies. The fire departments in Lancaster County are centrally dispatched by the County Fire Dispatch Center.

All of the fire departments are completely staffed by volunteers except for the City of Lancaster Fire Department, which is a career department. Lancaster County volunteer companies and Lancaster City respond into each other's jurisdictions using automatic mutual aid.

All fire apparatus is dispatched by the County Fire Dispatch Center, with the closest unit dispatched first. Each volunteer fire department determines the number and type of emergency units needed to respond within their district. Commercial buildings and public assembly occupancies are given box alarm assignments including up to four engines plus units such as aerial ladder trucks, tankers, and rescue squads. The box alarm for the theater fire in the Strasburg district called for three engines (Class A pumpers) and two tanker trucks. This assignment was established because Strasburg is a rural area without municipal water services, and tankers are needed to supply water for fire operations.

The County Firemen's Association employs a Fire Coordinator who coordinates the fire suppression, administration, and fire prevention efforts countywide. There is also an Assistant Coordinator who is primarily responsible for fire safety education efforts in the county. The Fire Coordinator and staff are primarily responsible to provide day-to-day administration of the Firemen's Association for the Lancaster County fire service. The Fire Coordinator does not have any direct operational authority on emergency incidents. However, the Fire Coordinator assists with incident command and other duties, such as public information and liaison.

The Firemen's Association is organized as a confederation of individual departments. Each department controls the training standards, operating procedures, and selection of operational and administrative officers. The Firemen's Association implements standard operating procedures for countywide operations as agreed upon by the members' consensus. These countywide procedures include centralized equipment dispatch, radio communications procedures, and emergency incident command protocols.

The Strasburg Fire Department, like all Lancaster County volunteer fire departments, is self-supported by fund drives and other fund-raising events, such as community dinners. Many of the departments enjoy good community support. They do not receive any direct funding from local or State governments, although the State government provides firefighting training that often is coordinated through the counties.

The Strasburg Fire Department requires completion of the State-certified Firefighter Essentials Class before members become full-fledged firefighters and are permitted to handle interior firefighting. New firefighters without State certification (i.e., junior or probationary firefighters) are allowed to respond to fires after completing company-level training on basic hose and ladder procedures. Junior firefighters without State training are limited to lower-risk outside duties. State training is supplemented by periodic company training sessions, which are the responsibility of the Captain, with oversight from the Chief.

Each company's membership elects its chief and line officers. Line officers receive additional training, such as incident command, from the Firemen's Association.

FIRE BUILDING

The fire occurred in a modem legitimate theater complex built specifically for live stage performances. This public assembly building had seating for 1,400 and was advertised as the largest indoor Christian theater in the nation. The theater specialized in presenting epic biblical dramas from a 100-foot main stage and two 75-foot side stages that formed a U-shape front, right, and left of the audience seating. The Sight and Sound Theater Complex included four buildings constructed at different times that were interconnected. This included prop and scenery construction and assembly buildings, prop and costume storage buildings, as well as specific carpentry and electrical shop areas (see Diagram 1). The complex also included stand-alone barns for the animals and livestock used in the performances.

The original stage and auditorium were built in 1990. The plans were reviewed by the Pennsylvania State Department of Labor and Industry in accordance with the Fire and Panic Act originally issued in 1927 and revised in 1984. The Fire and Panic Act cross references other State codes, including sections such as safety, fire prevention, egress, stairs, exit doors and exit access doors, and general fire alarm requirements.

The State Fire and Panic Act is a minimum code. Local governments are permitted to adopt stronger fire and building codes. The State's code does not incorporate or reference the model (standard) building, fire, or life safety codes.

The theater was built of steel rigid frame construction to allow for the large open space of the auditorium, unobstructed by columns. This construction typically employs columns along the walls that widen at the top and attach to the roof support girders. The columns are connected under the floor with turn buckles and steel rods. This type of construction requires wind bracing that usually con-

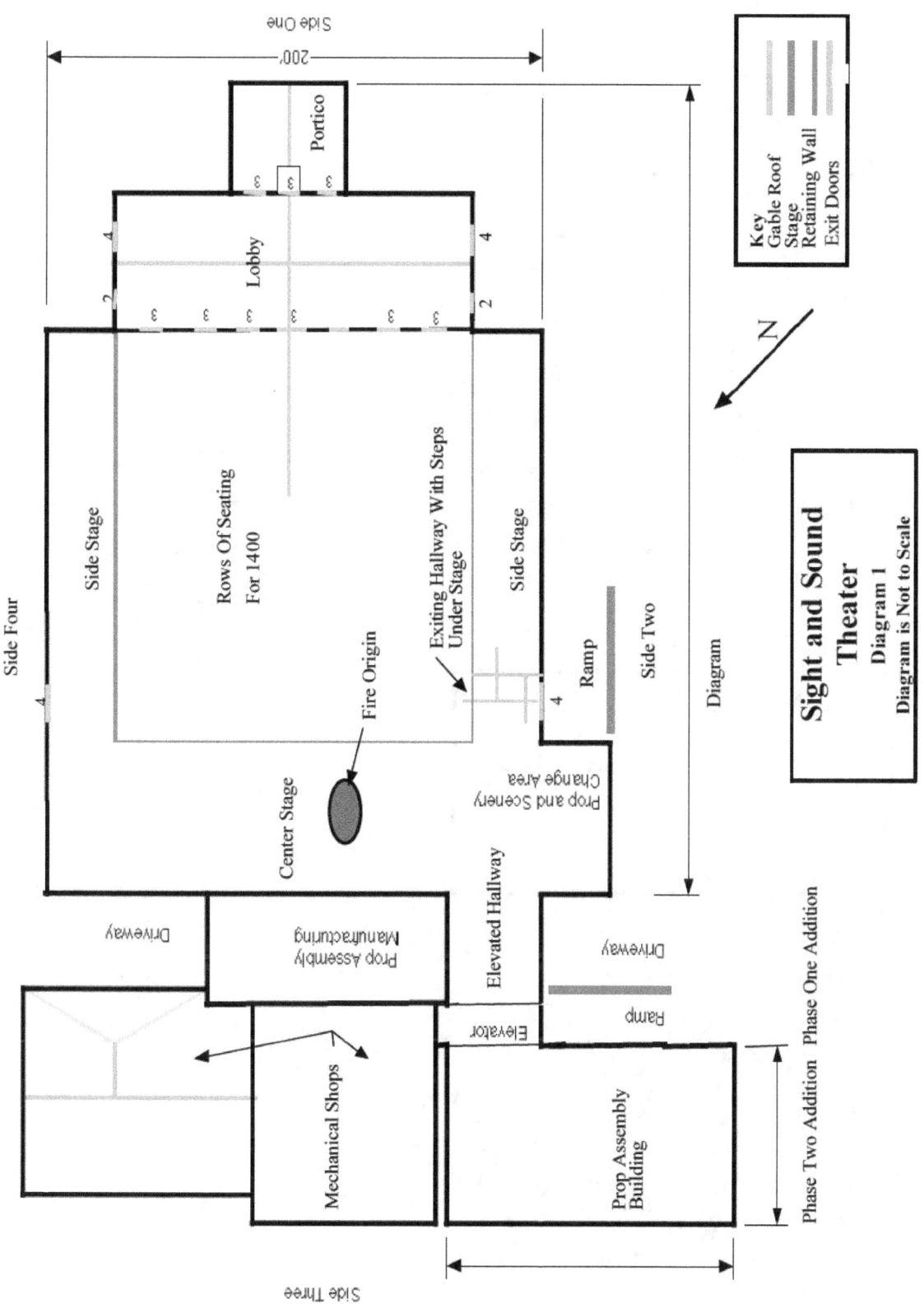

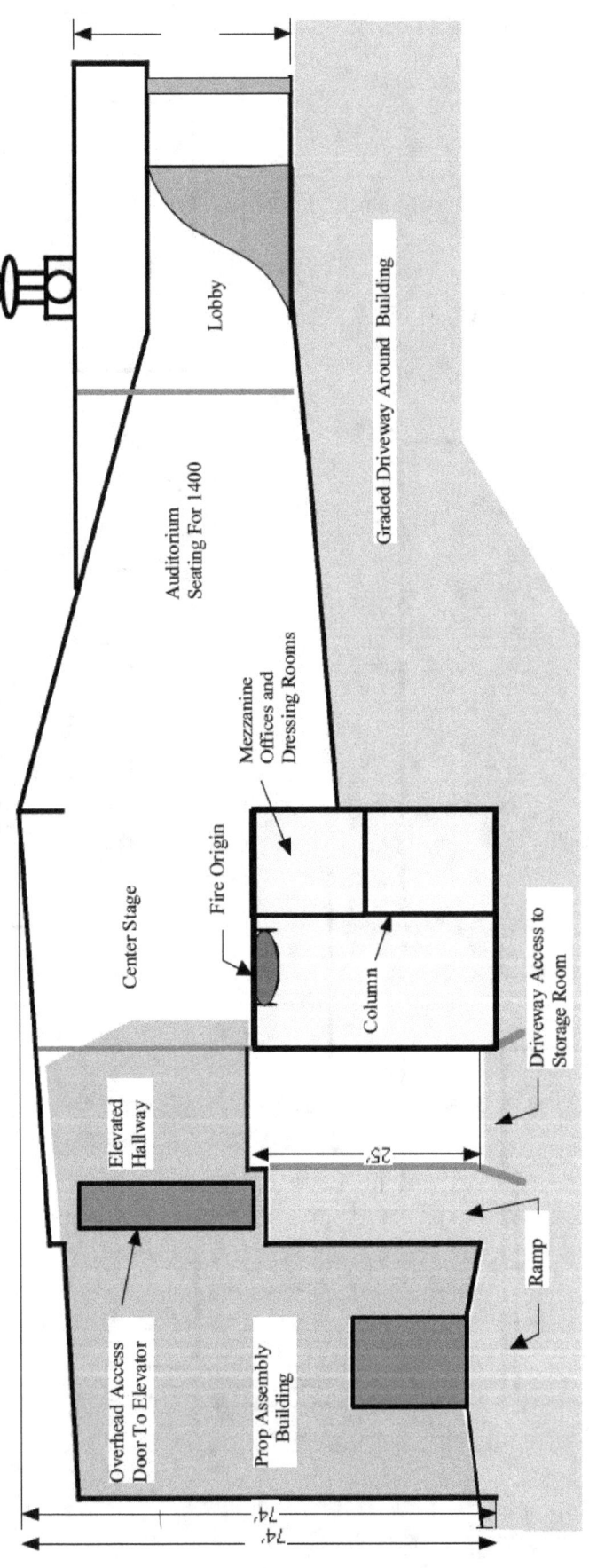

sists of diagonal steel rods and turn buckles between columns. The exterior walls had metal panels on the side and rear walls. There were no windows in the auditorium and only three windows in the rear of the side stages. The interior finish in the auditorium was drywall. The front entrance structure consisted of brick veneer walls, a gabled metal roof, and palladian windows in a colonial architecture design.

The storage area under the stage was accessible at grade level in the rear of the building. This room had a 25-foot ceiling with an elevated mezzanine for offices and dressing rooms accessed by a separate hallway leading from the stage. The storage area was accessible through a passage door and an overhead door. The storage room utilized the high ceiling space for tall props as well as the lower ceiling space (approximately 12 feet) under the mezzanine. The main electrical circuit boxes were also located in this area.

The Sight and Sound Theater grew through multi-phased additions. The first phase included a prop construction building of rigid steel with a built-in open shaft freight elevator. The elevator had 40-foot exterior roll-up doors to facilitate the movement of props and scenery into and out of the building. This addition was connected to the main stage by an elevated hallway approximately 40 feet wide by 40 feet high which did not have fire doors. The prop construction building was used to fabricate props and transport them by elevator through the elevated hallway to the main theater stage (see Diagram 2).

The second phase of construction for the theater complex was the prop assembly building. When completed, this addition connected the prop construction building to the new prop assembly area, and ultimately to the theater. The prop assembly building was built using a steel rigid frame and steel clad exterior and roof.

Exit and Life Safety Features

The auditorium appeared to meet acceptable standards for aisle and seating configuration and exiting[2]. It was equipped with proscenium curtains on the main and side stages, which are standard life-safety features in this type of building.

The 29 auditorium exits were adequate to quickly evacuate a typical audience, using a measure of 60 people per exit door per minute. However, the patrons of this theater tend to be elderly and more mobility-impaired than "typical" theater-goers. Had the theater been occupied, patrons using the eight exits closest to the stage would have had to move toward the smoke, and the exits on side two near the stage had steps which would have slowed egress.

Fire Protection Features

The Sight and Sound Theater was built in a rural township without tie-in to the municipal water supply. When the theater (Phase I) was built, a drafting pond was established behind the main building to provide water for fire suppression. As subsequent additions to the complex were initiated, the pond was eliminated to free space for new construction.

The Pennsylvania State Fire and Panic Act requires public assembly buildings with a capacity greater than 500 people and a storage room of over 100 square feet to have an automatic sprinkler system.

---

[2] The property owners would not permit USFA incident investigators inside the building during the site visit. The safety features below are based on a review of pictures taken at the same time, non-scaled building plans, and interviews.

The owner requested that the State Labor and Industry Department waive the sprinkler requirement because of the financial hardship caused by providing a water supply for the sprinkler system. The owner also argued that the possibility of accidental sprinkler discharge could cause costly damage to props and electrical equipment. The owner offered to provide a centrally monitored combination smoke and heat detection alarm system and to increase the fire resistance rating of structural assemblies in the storage room to two hours, both of which were done.

Once the theater was operational, the centrally monitored smoke detection system in the stage area was prone to false alarms due to theatrical smoke used on stage. These nuisance alarms interfered with stage performances and resulted in unnecessary fire department responses. After numerous fire department responses the theater management replaced the smoke detectors with heat detectors. The investigation conducted by the Pennsylvania State Police revealed that the theater management routinely shut off the alarm system during stage performances to prevent interference from false alarms. Instead, a fire watch was maintained using trained theater staff.

The stage storage area, prop assembly building, and prop maintenance building were protected with a sprayed-on fire resistant coating on all structural steel. The plans called for the coating to meet a two-hour fire resistance assembly rating. The sprayed-on coating, which was susceptible to damage from the movement of theater equipment, was protected by attaching plywood coverings on the columns to a height of eight feet.

The walls of the storage area beneath the stage were layered drywall to provide a two-hour fire protection rating for the mezzanine offices, and sprayed-on fire-resistant coatings on the structural steel columns and ceiling bar joists supporting the stage floor.

FIRE ORIGIN

The theater was scheduled to be closed from January 13th until March 8th, 1997, and was undergoing renovation work to make the stage floor more rigid. This involved removing the existing floor covering to expose the corrugated steel decking, and welding four foot by eight foot sheets of 1/4"-thick steel on top of the decking. The Pennsylvania State Police fire investigator who investigated the fire determined that the fire originated under the stage in the storage area. The fire was caused by a construction worker welding steel plates on the stage floor decking directly above the point of fire origin. During the removal of the floor covering, screw holes were exposed which allowed sparks and/or a molten arc welding rod to fall onto combustible props stored below.

The fire was discovered by two theater employees who went to the storage area for equipment and saw a stored stage prop on fire at three points: near the top of the prop almost at ceiling level, at waist level, and at the floor. All three fires were in a vertical alignment consistent with the showering of sparks from above.

The immediate reaction of the theater employees was to use nearby fire extinguishers to fight the fire. The first extinguisher was used on the two lowest fires. Dry chemical extinguishing agents normally work on ordinary combustibles by smothering the fire with a layering agent and inhibiting the fire's chemical chain reaction. On vertical surfaces the dry chemical agent may not adhere sufficiently to complete extinguishment. The powdery agent slides off, allowing re-ignition. Unable to completely put out the fire, both employees ran for additional extinguishers. At approximately the same time the welder smelled smoke but disregarded it, thinking it was the welding and the hot steel burning the soles of his boots. After the smell of smoke persisted, he located the foreman and they went below the stage to investigate. The construction employees arrived after the theater employees had used the second extinguisher.

The theater employees then left to call 9-1-1 and notify the other employees to evacuate the building. The two construction workers continued the attempt to fight the fire with portable extinguishers (approximately eight were used) until they realized they could not control the fire. They closed the exterior storage room door to limit the air supply, not knowing that another door from the auditorium to the storage room was blocked open by a fire extinguisher.

FIRE SPREAD

The two theater employees told the State Police Fire Investigator that when they first discovered the fire they noticed that the sprayed-on fire proofing had been knocked off the underside of the stage floor bar joists and support steel. The fire proofing was hanging on the wire mesh used to hold the coating to the overhead. The investigation revealed that the construction company's removal of the stage floor covering down to the corrugated decking involved striking the floor hard enough to knock off the sprayed-on protection, exposing the structural steel and bar joists in the storage area.

Another contributing factor to the fire spread in the below-stage storage area was on-going construction to the mezzanine rooms. Some walls were being changed and doorway cut-throughs were being added, but fire doors were not yet installed. This allowed fire to pass freely through these openings.

The fire spread vertically from the storage area to the stage, causing the steel to lose its tensile strength. Temperatures of 1,000° F can cause buckling and temperatures of 1,500° F can cause steel to lose strength and collapse[3]. When the heat and hot gases reached the stage ceiling they extended horizontally into the auditorium, causing the roof to fail all the way to the lobby fire wall. The fire also extended horizontally from the stage to the elevated hallway, causing the structural steel to fail and buckle in the prop assembly and prop maintenance buildings (see Diagram 3).

RESPONSE

The Lancaster County Fire Dispatch Center received a 9-1-1 telephone call from one of the Sight and Sound employees at 9:19 a.m. The first-alarm fire departments were dispatched for a fire in the building at 9:21 a.m. The initial assignment consisted of four engines and a tanker which was altered from the established assignment by the emergency equipment dispatcher because, from the 9-1-1 call, it was known that this incident was a working fire. The dispatcher added an engine to the assignment because the second tanker response would be delayed.

Since the Strasburg Fire Department had responded to many needless alarms at the Sight and Sound Theater, the Dispatch Center alerted the responding units that this incident was not initiated by the alarm system. Strasburg Volunteer Fire Chief Rick Wentz immediately requested that a ladder truck be added to the assignment.

The first Strasburg engine to arrive at the theater dropped a 5-inch supply line at Hartman's Bridge Road and laid out 800 feet of hose up the driveway. There, theater employees signaled where to position the engine to access the fire. The Strasburg engine set up in the parking lot near the stage on the east side of the building. There was no fire visible because of the few windows and openings in the building.

---

[3] Branigan pg. 232

Theater employees advised the officer on the first engine that all employees were out of the building and were accounted for in the parking lot. The employees then directed the firefighters to the storage room where they had discovered the fire.

Fire Response

The Strasburg Chief responded in his vehicle and used it as the command post.

Fourteen volunteer fire departments responded to the Sight and Sound fire committing the following resources:

> 25 engines
> 4 aerial ladder trucks
> 22 tankers
> 5 rescue squads
> 2 air re-fill units
> 2 emergency medical units

The specific departments involved were as follows:

- Lampeter
- Bird in Hand
- Paradise
- West Willow
- Lafayette
- New Danville
- Shawnee
- Eden
- Quarryville
- Willow Street
- Ronk
- Lancaster Township
- Witmer

Fireground Organization

Strasburg Chief Wentz arrived immediately after the Strasburg units and called for a multiple alarm. He directed units to establish a water supply relay using 5-inch hose laid from a pond a quarter mile away behind a nursery (visible in aerial view two). The Chief ordered the responding ladder trucks to attempt to ventilate the theater roof, and had additional attack lines advanced into the auditorium near the stage.

The Lancaster County Fire Chiefs Association has adopted a standard operating procedure for incident command. The incident command procedure identifies each side of the building by number and establishes the responsibilities of the incident commander and sector officers. (For additional details see Appendix A.) Chief Wentz established incident command at his vehicle and made the following assignments:

COMMAND OFFICER	ASSIGNMENT
Chief 4-1	Water supply sector
Deputy Chief 5-4	Side One sector (front of building)
Deputy Chief 6-3	Side Two sector (west side)
Chief 5-5	Side Three sector (rear of building)
Deputy Chief 5-1	Side Four sector (east side)

Public Information Officer functions were assigned to the Lancaster County Fire Coordinator.

Much of the staffing and resources were directed toward establishing a water supply. This required three engines drafting from the pond, with additional engines pumping in-line to complete the lengthy 5-inch hose relay. Tankers were also used to shuttle water to the theater parking lot.

The demand on fire department resources required a third alarm. Apparatus staging was managed in the parking lot and a firefighter rehabilitation area was established. Tactical radio frequencies were assigned to water supply, EMS operations, ladder truck operations, and general fireground operations.

The firefighters advanced a 1-1/2-inch line with a fog nozzle to the rear of the building at the storage room passage door. Firefighters were unable to advance into the room because the entranceway was blocked by storage props, sets, and other large items. The firefighters then entered through the overhead door and advanced into thick black smoke (see Diagram 3). They advanced into the storage room but were unable to see the fire and were reluctant to apply water onto smoke without visible flame. As the operation progressed, the initial attack crew in the storage room was relieved by another crew when their SCBA became depleted. The crew on the attack line in the auditorium encountered untenable heat and zero visibility. Also of concern was a large stationary liquid nitrogen tank at the outside wall of the storage room. A hose line was dedicated to keeping the tank cooled to prevent overheating.

Once the heat of the fire caused the structural steel to fail in the storage area (aided by the damage to the sprayed-on fire protection during renovation), interior firefighting became too hazardous to continue. The truck crews ventilating the roof noted metal discoloration and buckling steel. The sector officer for that area noted exterior heat discoloration extending in the auditorium and the prop assembly building, and the order was given to evacuate the building and discontinue interior operations. As a result, the fire department adopted defensive tactics and re-positioned units for an exterior attack. Only after the roof collapsed and water from exterior streams could reach the fire did the fire department begin to make progress at extinguishing the blaze.

Defensive Operations

With all fire department personnel evacuated from the buildings, several ladder trucks were repositioned for ladder pipe operations. It was estimated that the master streams on the building were flowing 5,400 gallons per minute at the peak of the fire.

The master streams were not immediately effective because there were very few building windows or other suitable openings through which to direct the water. The theater had exit doors in the auditorium; however, these openings were too low for effective exterior master stream operations. Portable master stream devices can not be operated safely through doorway openings in close proximity to walls that could collapse. Once the auditorium's roof collapsed, water streams were able to reach the fire. The fire was declared under control at 1:02 p.m., approximately 3 1/2 hours after the initial alarm.

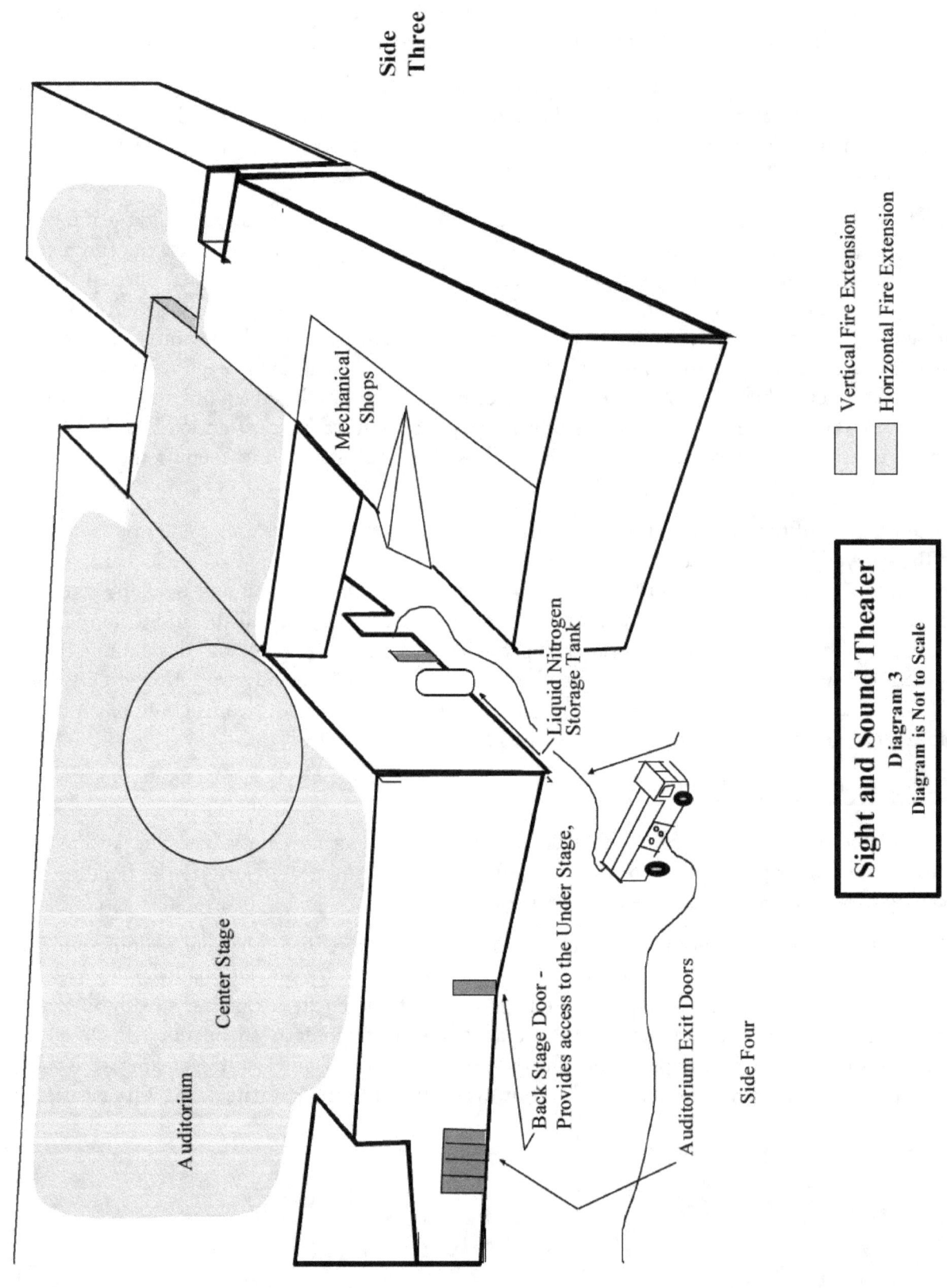

Sight and Sound Theater
Diagram 3
Diagram is Not to Scale

Aerial View One – shows the fire in progress after the auditorium roof collapsed. Note the four master streams and the tanker relay hose lines.

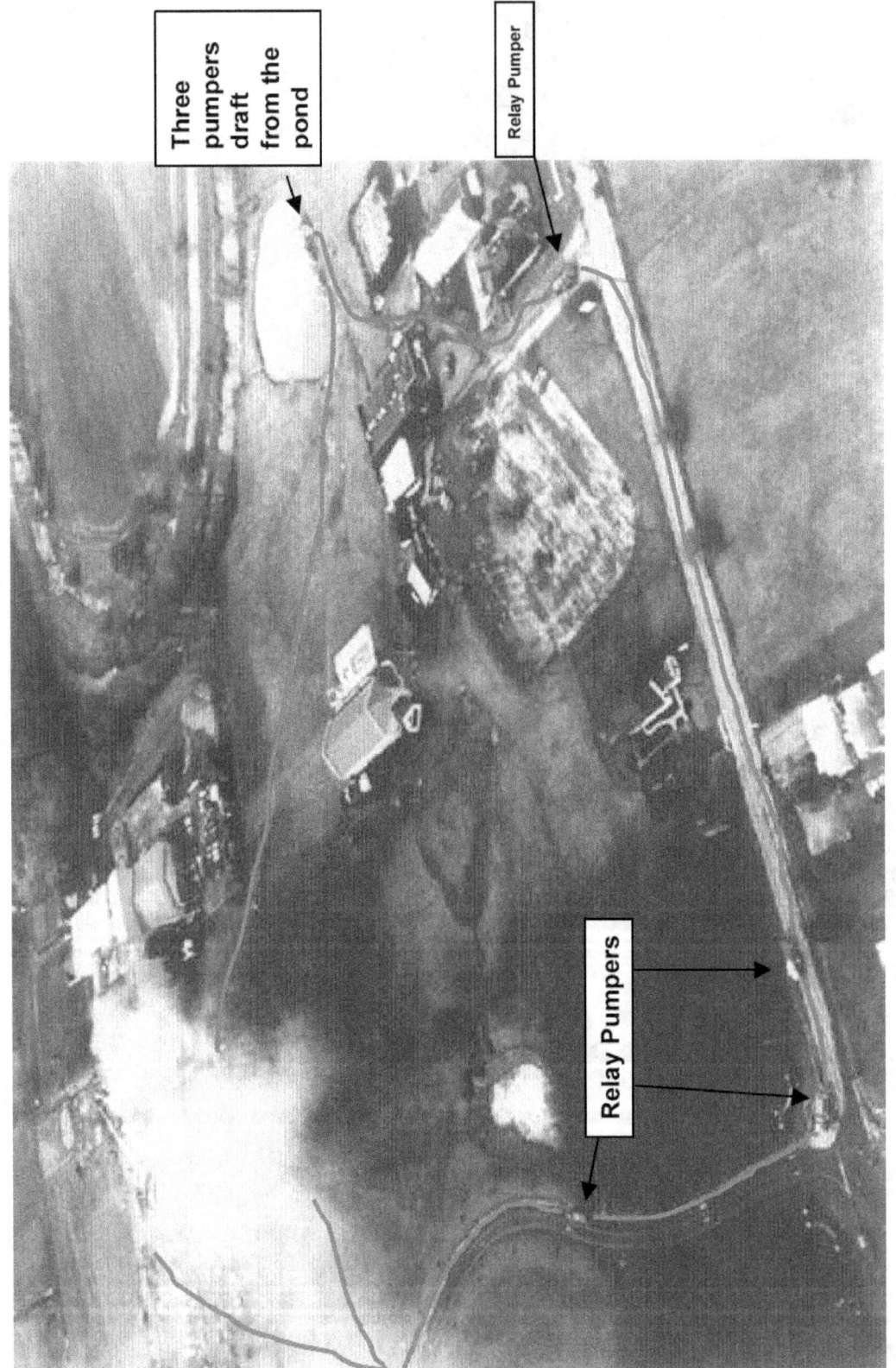

Aerial View Two—shows the water supply relay system established at the fifth fire water pond, which is the approximate location quarter mile trail from the fire. Three pumpers at the scene indicate the area obscured by the black smoke.

This picture is taken from the front, looking toward side two of the building, which shows the auditorium roof and side stage roof collapse.

The front of the theater side one shows little damage because the fire wall between the auditorium and lobby held.

Shows the stage floor corrugated decking and the steel plate renovation in progress. Also shows structural steel failure on the main stage.

This picture was shot from the direction of the stage showing the roof and structural steel failure in the auditorium. The heat level is indicated by the damage to the seating

This is the skeleton of the 40 x 40 foot hallway which allowed the fire to communicate to the prop assembly and maintenance buildings.

Shows the failure of the rear stage separation wall between the mechanical shops.

Visible here is buckled steel from the wall and roof of the prop maintenance building. The heat traveled horizontally from the stage roof through the 40-foot hallway to the building.

The combustible fuel load in this building was relatively low in comparison to the storage area. The once sprayed on fire protection only remains on the lower portion of the vertical column.

Remains of some scenery prop frames which were obstructions to hose line advancement and locating the fire. Also visible hanging from the ceiling is the wire mesh used to support the spray-on fire protection which was damaged during the floor renovation.

Shows the under stage storage room and point of fire origin. Visible are buckled columns and the mezzanine level.

ANALYSIS

This fire is notable for the breakdown of the systems originally intended to provide basic fire safety in the Sight and Sound Theater.

In 1989 while the building was in the design process, the Strasburg Fire Chief cited the need for automatic fire sprinklers, an access road to the nearby pond for water supply, and a rooftop ventilation system to protect the theater. The fire department's concerns were not afforded adequate attention by the property owners, nor some State Department of Industry and Labor personnel.

The owners were granted a waiver for the required fire sprinkler system in the storage area (where the fire started), and were required instead to provide an automatic, centrally monitored alarm system. The rationale for the request and the justification for the approval was explained by John J. McNulty, Chairman of the State Industrial Board as reported by Stephen Tropnell of *The Lancaster New Era* newspaper, January 30, 1997. McNulty said that alternate fire protection was approved for several reasons:

- The potential damage to electronic equipment, costumes, sets, and props by accidental discharge of a sprinkler would be substantial, possibly hundreds of thousands of dollars.

- The entire building was constructed of non-combustible materials.

- The storage room where sprinklers were required was constructed of materials with a two-hour fire resistance rating.

- Obtaining a reliable water source for the sprinkler system would be very difficult.

The State Department of Labor and Industry reasoned that the building was built of non-combustible materials and that theatrical pyrotechnics and smoke used during performances could accidentally activate sprinkler systems, potentially causing water damage.

The automatic centrally monitored alarm system included smoke and heat detectors. However, the smoke detectors on stage were often activated by theatrical smoke which set off needless alarms. This led to the conversion of smoke detectors to heat detectors and started the practice of disabling the system during performances, and using employees as a fire watch.

The lessons of the Sight and Sound fire are not new. This fire is very similar to other fires in history, examples are:

- 1967, Chicago, Ill., McCormick Place exhibition hall fire where a non-sprinkled, fire resistive, protected steel constructed building collapsed. This fire was not occupied with patrons at the time of the fire. Maintenance crews tried unsuccessfully to control the fire, and the notification to the fire dispatcher was delayed. The fire protection water system maintained by the property owner was not in operation. The fire was fast burning and had high heat production which contributed to the steel failure and total collapse.

- 1985, Fairfax, VA., Wolftrap Farm Pavilion fire began in the stage at the prop storage area and resulted in total destruction. Recommendations for sprinkler protection were not heeded. This theater was set back off the main road and was closed when the fire started. Late discovery allowed the fire to gain much headway before the fire units arrived.

- 1904, Chicago, Ill., Iroquois Theater fire. This theater was occupied when a fire occurred in the stage scenery. The fire rapidly extended from the stage to the auditorium. The death total was 602 people who could not get out before they were overcome from toxic smoke.

In the case of the Sight and Sound fire the following factors contributed to the loss:

1. Fire department notification was delayed because the alarm system did not activate. The alarm company received no signal for an alarm at the theater[1].

2. The two-hour fire-resistance-rated assembly in the storage area beneath the stage was damaged during the stage floor renovation, leaving the structural members unprotected from the ensuing fire.

3. Even though the importance of water supply was known by the property owners (lack of it was cited as a reason for not installing sprinklers), the pond initially provided to supply water for fire suppression was subsequently eliminated to make room for a large addition. The recommended access road to the alternate water supply pond was not built because the owner of that land would not allow the construction. With no access road to the off-site pond, fire apparatus became stuck in the mud and had to be towed. This also limited access for additional units trying to establish water supply.

4. The props assembly and maintenance buildings were connected to the backwall of the theater stage with a 40 foot by 40 foot hallway which had no fire separation door. The State Fire and Panic Code, Subsection 51.2 (see Appendix B) requires that when public assembly occupancies share a structure with other occupancies, the other structure will be separated with fire walls or be governed by the most restrictive limitations. The lack of sprinklers and fire separations contributed to fire extension and damage to the prop assembly and maintenance buildings.

5. The representatives from the State Department of Labor and Industry did not apparently understand how a two hour fire resistance rating should be applied per the particular features of a specific structure. The level of protection is variable depending upon many factors that affect the rate and extension of heat and flame. The two hour rating does not guarantee two hours of fire resistance for buildings. Rather, the rating relates to a laboratory test conducted using a controlled time temperature curve where an assembly is exposed to increasing heat levels until 1,850° F is reached at two hours[2]. In real fire situations, fire temperatures can reach 1,850° F in minutes, not hours.

   Misunderstanding of the fire resistance rating resulted in a mistaken justification for waiving the required storage room sprinkler system. Storage rooms often have high fuel loads which, when involved in fire, produce enough heat to quickly overwhelm the fire resistant design intended to contain the fire within a compartmentalized area. The code recognizes storage areas in public assembly occupancies as potential fire hazard areas that require dynamic protection features to extinguish or contain a fire.

   The two hour fire resistance assembly test is consistent with the combustion of ordinary Class A materials capable of heat generation at 8,000 BTUs per pound. The under-stage storage area in the theater held props constructed of plastic resin glass fibers, polystyrene foam, and plastics that typically produce heat generation rates of 16,000 BTUs per pound, twice the rate of ordinary combustibles. In addition, the canvas, cloth, and wood used in the props had high surface area to mass ratios conducive to rapid flame spread. The heavy fuel load in this facility would be

---

[1] USFA incident reviewers were not allowed to examine the system.

[2] NFPA Fire Protection Handbook, section 5, pg. 64

considered by most life safety and fire codes to be hazardous storage, thus requiring sprinkler protection.

6. The stage floor renovations contributed to the overstocked storage room and heavy fuel load. The theater routinely produced four separate seasonal productions throughout the year. Normally, three productions were in storage while one remained on stage. Due to the stage floor construction, the props and scenery for the fourth production also had to be crowded into the understage storage area. This further increased the fuel load and impeded the advancement of hose lines by firefighters.

7. The inability to efficiently ventilate smoke from the theater further complicated the efforts of firefighters to find the vertical fire travel in the auditorium. The inability to effectively relieve the heat vertically through the roof allowed the intense thermal effects to travel horizontally across the underside of the roof, causing more damage to the steel structure and the subsequent roof collapse. The automatic rooftop ventilation originally suggested would have mitigated the heat and smoke extension dramatically.

8. Although there were six civilians who suffered minor injuries, the fire suppression operation was conducted without firefighter injury. This fire presented a significant potential for firefighter injuries during the interior fire attack phase and from the subsequent roof collapse. The fireground command officers recognized the imminent potential for building collapse, and removed personnel from the structure. They then used defensive tactics and conducted operations from the exterior.

9. The theater auditorium was designed to allow for the evacuation of 1,400 occupants through the 26 exit doors without steps. The exiting rate of 60 people per minute is based upon healthy, ambulatory people[3]. The Sight and Sound Theater productions were family programs targeting young children and elderly audiences. As much as 60 percent of the typical audiences consisted of elderly individuals or children who may have required exiting assistance. Many of the elderly who usually attended used wheel chairs. These circumstances almost certainly would have delayed exiting to a rate of fewer than 60 people per minute had the fire occurred while the theater was occupied. Also, smoke and the products of combustion probably would have communicated into the auditorium, contributing to additional exiting problems.

LESSONS LEARNED

1. **Local governments have the responsibility and fire departments need the authority to ensure that local fire protection capabilities can cope with a fire of any structure, especially a high occupancy public assembly such as the Sight and Sound Theater.**

    Unfortunately, many volunteer fire companies across the country are not provided a formal voice into the building plan process. Where this is the case, there are several actions that fire departments might otherwise take if faced with new fire protection demands that exceed the staffing, equipment, and training resources of the department, especially when built-in protection requirements appear to be compromised:

    A. Develop a process whereby the fire department is apprised of building permit requests and plans.

---

[3] NFPA Fire Protection Handbook, section 6, pg. 9

B. Notify the local government if the proposed construction places excessive demands for fire and life safety protection. Provide formal recommendations, in writing, for built-in fire protection and water supply requirements as appropriate.

C. When waivers are requested which would adversely impact fire protection coverage, fire department concerns should be communicated in writing to the appropriate local and State agencies.

D. If structures are built without adequate fire protection features and are beyond the resources of the local fire service to protect, the local fire service organization could notify the property owners, local government, and the property's insurance carrier that major property loss will likely occur beyond what would be expected had the structure been built code compliant. The fire department could also point out that the planned construction poses a potential life safety risk to both the firefighters and general public.

E. The local fire service organization could work toward the adoption of appropriate local building and fire codes with local authorities.

2. **Pre-construction meetings should be conducted with property managers and contractors, so that fire service representatives can provide important insight for fire safety design and fire protection measures.**

On complex buildings and public assembly occupancies, pre-construction meetings allow the fire department representative to review the site and building plans and make fire protection recommendations. The plans review process normally allows the fire department to offer valuable information to the property owners and contractors, and enables the department to become familiar with how the planned construction could affect fire protection systems and influence fireground operations. Such conferences should be also held prior to any major additions or repairs to existing structures.

If a meeting had been held with the local fire department before work proceeded on the stage floor at the Sight and Sound Theater, the fire department representative would likely have recognized the potential hazard of welding over the storage area, and could have suggested fire prevention measures, such as a fire watch during welding operations, and removing or protecting the combustibles below the work area.

3. **Government and fire code officials must recognize that building safety/fire code requirements are usually based on past tragedies. Waivers and tradeoffs of code provisions must be granted sparingly and only with adequate, comparable fire protection coverage through other means. Otherwise, the safety of citizens and firefighters is jeopardized.**

Public assembly occupancies warrant more stringent requirements because of the high potential for multiple casualties and large dollar losses. Claims of economic hardship caused by complying with codes rarely overshadow the cost of the fire that otherwise might have been prevented or suppressed in the incipient stage. This fire is a perfect example. The cost of the fire is estimated at $15 million dollars lost to the structure, $55 million dollars lost to the tourist industry, thousands of dollars in response costs expended by the fire departments, and the loss of jobs for 62 theater employees. (See Appendix C.) The cost associated with the fire protection features that were $225,000 pales in comparison to the subsequent loss.

Fire protection is a service provided by people from the community and should be supported by the community. This support should include responsible construction of buildings which meet life and safety codes and do not exceed the capability of local resources.

4. Buildings constructed of steel should, in effect, be considered unprotected and capable of collapse from fire in as few as ten minutes.

Fire resistant coatings sprayed onto structural steel are susceptible to damage from construction work. As a result of the delayed call to 9-1-1, the lack of a sprinkler system in the storage area, and the failure of the alarm system, the fire in the storage room beneath the stage was probably already unsafe for an interior attack when the 9-1-1 call was made.

Building construction training is especially important for line fire officers. Basic fire training does not generally cover building construction or pre-fire planning. All fire officers need to be familiar with the specific hazards of building construction. Understanding the risks associated with building collapse will increase safety of firefighting operations.

5. The impact of fire and heat on structural steel members warrants extreme caution by firefighters.

For a fire of this magnitude the interior attack should be at least two 2-1/2-inch hose lines-- one 2-1/2-inch hose line for fire attack and the second as a back-up line to direct water on the structural overhead. For reach and penetration at least one line should have been equipped with a smooth bore nozzle. Water directed on heat-buckled or sagging steel members generally will prevent further distortion of the steel, provided the fire is being extinguished.

Although the first arriving engine did not run out of tank water before the 5-inch hose relay was complete, the use of four 2-1/2-inch lines would have required approximately 1,000 GPM.

6. The general firefighting rule of using water only on visible fire (not on smoke) is not applicable to structural steel construction.

Unless the steel members are cooled with high-volume hose streams, the fire's heat can rapidly cause steel to lose its strength and contribute to building collapse. Therefore, interior fire attack in steel construction should include the timely use of sufficiently sized handlines to cool the structural steel, with additional lines deployed to attack the fire.

7. A personnel accountability system should be implemented and strictly applied by a designated safety or accountability officer, on every incident and particularly at working structure fires.

While there were no significant injuries, a safety officer should have been appointed by the incident commander. An incident this large with the potential for firefighter injuries warrants a personnel accountability/safety manager. This position is needed to coordinate the overall safety of fireground operations and to ensure personnel accountability.

The establishment of effective accountability systems for all personnel operating at the scene of fires has become a standard safety practice in the fire service. This type of system can greatly reduce the risk of over looking personnel during the evacuation of the fire building. It also reduces the possibility that personnel may become trapped or incapacitated without the Incident Commander being aware of their presence. Although many fire departments (see Appendix D for Lancaster County's) have accountability procedures in place, it is important that they be implemented routinely.

8. **Once incident command is established, it is valuable to assess whether sectors should be assigned by geographic position or by tactical function.**

   With the exception of water supply, sectors were assigned geographically based on building sides. Consideration should be given to tactical areas of responsibility such as ventilation, exposure protection, planning, safety, etc. Tactical sectors are often more effective on fires of this magnitude because the sector commanders can better coordinate specific functions for the entire incident. The ventilation of the structure was critical because of the limited openings and the difficulty in locating the fire. If the large overhead doors at the prop assembly building could have been opened, it would have relieved the heat at exposure number three. A hose crew was employed on the liquid nitrogen tank at the storage room entrance. The planning sector, if utilized, could have concentrated on how to open the overhead doors, and the best way to handle the nitrogen tank hazard.

9. **Effective incident command of a major incident like this fire would benefit from a specialized command vehicle to support incident management functions.**

   On large-scale fires, a command unit with space for command officers to meet and have adequate communications capabilities is very important. The Lancaster County Fire Service has approximately 100 fire departments that could utilize a command vehicle for more efficient operations at major incidents. A critique of this incident held by the Strasburg VFD identified the need for a command bus on all multiple alarm incidents in the County.

10. **Codes pertaining to exiting in public assembly occupancies do not account for public assembly occupancies frequented by a high percentage of elderly and children.**

    The code requirements on building evacuations generally are established on the basis of healthy ambulatory people. Code provisions do not routinely consider the exiting impact of those needing assistance, such as elderly and handicapped people, and very young children. What are normal and acceptable exiting rates change when occupants are less ambulatory, especially when fire protection systems are waived.

    For the fire safety of public assembly occupants it is necessary that exits are adequate for timely evacuation, and that fire protection systems control the fire until evacuation is complete. A fire protection sprinkler system and a reliable alarm system are essential for ensuring the necessary evacuation time. This theater attracted large audiences which would have needed assistance during emergency exiting, and therefore, more time to evacuate the auditorium. Waiving required fire safety systems further increased the risks for fire- and panic-related injuries had the theater been occupied when the fire occurred.

BIBLIOGRAPHY

NFPA 101, Code for Safety to Life from Fire in Buildings and Structures. National Fire Protection Association. 1991. Quincy, Massachusetts.

NFPA Fire Protection Handbook. 15th Edition. National Fire Protection Association. 1981. Quincy, Massachusetts.

Branigan, Francis L. Building Construction for the Fire Service. National Fire Protection Association. 1971. Quincy, Massachusetts.

APPENDIX A

Lancaster County Pennsylvania

Incident Command System Fire/Emergency Standard Operating Procedures

Approved: Lancaster County Fire Chiefs' Association
December 12, 1990

INDEX

1. Purpose
2. Communications
3. Assumption of Command
4. Selection of Command Mode
5. Responsibilities of the Incident Commander
6. Standard Geographic Designation System
7. ICS Organizational Structure for Initial Operations
8. Sectors
9. Staging
10. ICS Organization For Larger Incidents
11. Transfer of Command
12. ICS Charts

1. PURPOSE

This procedure is established to:

1. Provide for the safety of personnel operating at emergency incidents through improved command and control (for management of emergencies).

2. Improve the use of resources and tactical effectiveness.

3. Comply with Occupational Safety and Health Administration's (OSHA) Hazardous Waste Operations and Emergency Response (29 CFR 1910.120) for hazardous materials incidents.

4. Meet National Fire Protection Association (NFPA) Standard 1500 and 1561 for the use of an Incident Command System for operations at all emergency incidents.

To meet the goals of this procedure: The Incident Command System shall be implemented at all incidents for which the fire service has management responsibility.

2. COMMUNICATIONS

ALL COMMUNICATIONS SHALL BE CLEAR TEXT.

Radio communications shall be received from sender using the following model:

1. Request to initiate communications and determine that the intended receiver is listening.

2. Transmit the message or order concisely in clear text.

3. Receive feedback from the receiver to ensure that the message was received and understood.

4. Confirm that the message or order was understood; If not, correct and clarify the message.

5. The Incident Commander is responsible for all fire communications.

Example of Clear Text Communications:

Sender	Message
Command 5-12:	Engine 5-12, Command
Engine 5-12:	Command 5-12, Engine 5-12
Command 5-12:	Protect exposure, Side 2
Engine 5-12:	Protect exposure, Side 2
Command 5-12:	Affirmative

Sender	Message
Command 5-12:	Ladder 5-12, Command 5-12
Ladder 5-12:	Command 5-12, Ladder 5-12
Command 5-12:	Establish vertical ventilation, access from Side 2
Ladder 5-12	Establish vertical ventilation, access from Side 3
Command 5-12	Negative, Establish vertical ventilation, access from Side 2
Ladder 5-12	Vertical Ventilation, Access from Side 2
Command 5-12	Affirmative

3. ASSUMPTION OF COMMAND

Command shall be established at all incidents.

The senior member of the first arriving Company is Incident Commander until relieved by more Senior Fire Official. When multiple resources will be committed to the incident, command shall be formally established by transmitting a brief initial report containing the following information to the Communications Center:

1. Identity of the Company transmitting the report.
2. Actual location of the incident.
3. Brief description of the incident and report of conditions. (Size Up)
4. Designation of the individual assuming command and incident name (if required).

Incidents are given a specific name to reduce confusion when multiple incidents share the same radio frequency and/or dispatcher.

Example:

Sender	Message
Engine 5-12:	County, Engine 5-12
County Comm.:	Engine 5-12, County
Engine 5-12:	Engine 5-12 on scene, 100 Willow St. Pike, fire showing from 1st floor, side 1 of a 2 story dwelling, Captain 5-1 is Command
County Comm.:	Engine 5-12 on scene, 100 Willow St. pike, fire showing from 1st floor, side 1 of a 2 story dwelling, Captain 5-12 is Command
Command 5-12:	Affirmative

4. SELECTION OF COMMAND MODE

The Incident Commander (IC) should conduct the initial command activity from a fixed position, particularly where an incident is escalating rapidly or complex.

If there is a need for immediate tactical activity, and company manpower necessitates that the IC be an integral part of tactical operations, command in the offensive mode should be initiated.

Command in the offensive mode should only be performed until command can be transferred.

5. RESPONSIBILITIES OF THE INCIDENT COMMANDER

The Incident Commander at any fire/emergency incident shall be responsible for the following:

1. Assessment of Incident Priorities. Incident priorities provide a framework for command decision making. Tactical activity may address more than one incident priority simultaneously.

    - Life Safety (first priority)
    - Incident Stabilization (second priority)
    - Property Conservation (third priority)

2. Perform Sizeup. The IC must perform an initial assessment of the situation, incident potential, and resource status. This assessment must address the following three questions:

   - What have I got? (situation)
   - Where is it going? (potential)
   - What do I need to control it? (resources)

3. Select the Strategic Mode. A critical decision having an impact on the safety of personnel and the effectiveness of tactical operations is the selection of strategic mode. Operations may be conducted in either an Offensive or Defensive mode. This decision is based on answers to the following two questions:

   - Is it safe to conduct offensive operations?
   - Is resource capability (present and projected) adequate for offensive operations to control the incident?

4. Define Strategic Goals. Strategic goals define the overall plan that will be used to control the incident. Strategic goals are broad in nature and are achieved by the completion of tactical objectives. Strategic goals are generally focused in the following areas:

   - Search, protection and removal of trapped/exposed persons. (Rescue)
   - Confinement and extinguishment of the fire or control of the hazard. (Fire attack or Containment)
   - Minimize loss to involved or exposed property. (Salvage and Overhaul)

5. RESPONSIBILITIES OF THE INCIDENT COMMANDER

The Incident Commander at any fire/emergency incident shall be responsible for the following:

1. Assessment of Incident Priorities. Incident priorities provide a framework for command decision making. Tactical activity may address more than one incident priority simultaneously.

   - Life Safety (first priority)
   - Incident Stabilization (second priority)
   - Property Conservation (third priority)

2. Perform Sizeup. The IC must perform an initial assessment of the situation, incident potential, and resource status. This assessment must address the following three questions:

   - What have I got? (situation)
   - Where is it going? (potential)
   - What do I need to control it? (resources)

3. Select the Strategic Mode. A critical decision having an impact on the safety of personnel and the effectiveness of tactical operations is the selection of strategic mode. Operations may be conducted in either an Offensive or Defensive mode. This decision is based on answers to the following two questions:

- Is it safe to conduct offensive operations?
- Is resource capability (present and projected) adequate for offensive operations to control the incident?

4. Define Strategic Goals. Strategic goals define the overall plan that will be used to control the incident. Strategic goals are broad in nature and are achieved by the completion of tactical objectives. Strategic goals are generally focused in the following areas:

   - Search, protection and removal of trapped/exposed persons. (Rescue)
   - Confinement and extinguishment of the fire or control of the hazard. (Fire attack or Containment)
   - Minimize loss to involved or exposed property. (Salvage and Overhaul)

5. Establish Tactical Objectives. Tactical objectives are the specific operations that must be accomplished to achieve strategic goals. Tactical objectives must be both specific and measurable, defining:

   - Assignment of resources
   - Nature of the tactical activity
   - Location in which the tactical activity must be performed
   - It the tactical action must be performed in sequence or coordinated with any other tactical action

6. Implement the Action Plan.

   Implementation of the incident action plan requires that the IC establish an appropriate organizational structure to manage the required resources and communicate the tactical objectives.

   The incident action plan may be communicated by your Standard Operating Procedure, assigning tactical objectives, or by assigning task activity.

   Tactical Standard Operating Procedures may define common components of the incident action plan such as water supply, standard apparatus placement, and the methods used for basic tactical evolutions.

7. Incident Safety.

   Incident Scene must be controlled to protect fire/emergency personnel and keep unauthorized persons out of hazardous areas.

   IC may delegate incident safety authority to an appointed Safety Officer who may cease or stop an operation without going through the Chain-of-Command, only where personnel are in imminent danger of being injured or killed.

   All Fire Officers must maintain a constant awareness of the position and function of all personnel assigned to operate under their supervision.

   IC must establish a personnel identification system to identify and keep track of personnel entering and leaving hazardous areas or areas where special protective equipment is required.

All Fire Officers must maintain an awareness of the condition of personnel operating within their control and ensure that adequate steps are taken to provide for their safety and health. This includes medical evaluation, food and fluid replacement, relief and reassignment of fatigued crews.

Orders from the IC may specify tactical objectives assigned to subordinate positions within the ICS structure or to a specific resource.

Example:

Sender	Message
Command 5-12	Engine 5-12, Command 5-12
Engine 5-12	Command 5-12, Engine 5-12
Command 5-12	Initiate fire attack on 1st floor as soon as Ladder 5-12 establishes vertical ventilation.
Engine 5-12	Initiate fire attack on 1st floor as soon as Ladder 5-12 establishes vertical ventilation.
Command 5-12:	Affirmative

6. STANDARD GEOGRAPHIC DESIGNATION SYSTEM

Each exterior of a structure shall be given a number designation using the term "Side". The side of the structure facing the street (address side) shall be designated 1. The remaining sides shall be designated 2, 3, 4 in a clockwise manner. Exposure shall be designated in a like manner as shown below:

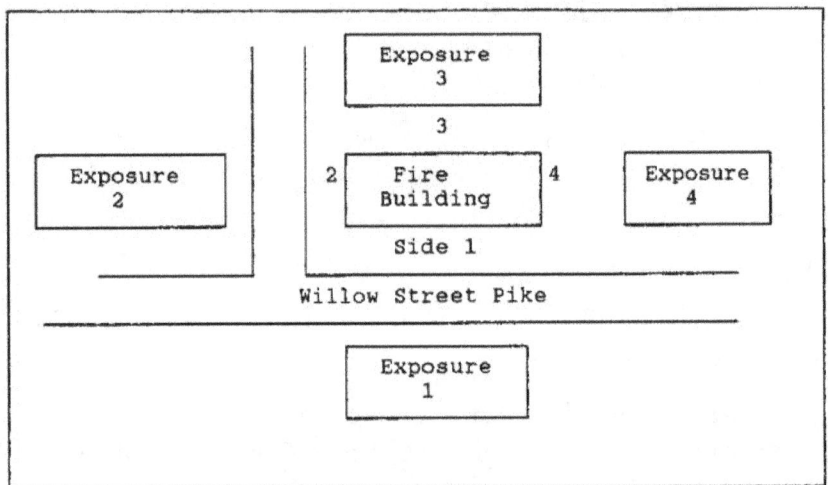

The interior of a structure shall be designated by the term "Floor" (1, 2, 3, etc.) and may be divided into quarters using A, B, C, D and E as shown below. The basement, attic, and roof shall be designated by name.

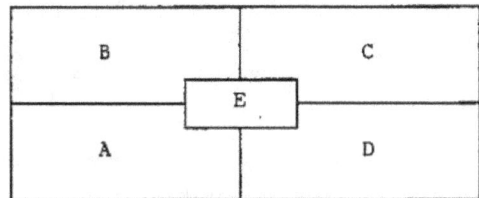

7. ICS ORGANIZATIONAL STRUCTURE FOR INITIAL OPERATIONS

The ICS shall be used to maintain an effective span of control and workload for all supervisory personnel. Each supervisor can effectively manage 3 to 7 personnel but ideally he or she should only manage 5.

Example:

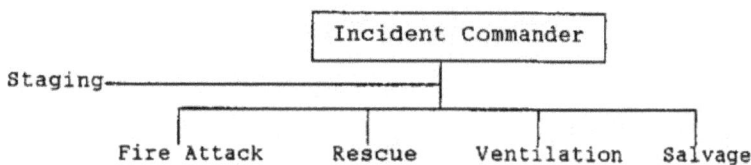

8. SECTORS

When multiple resources are assigned to the same function incident-wide (such as ventilation or search and rescue), a *"Sector Group"* shall be established to provide coordination and control of tactical operations.

When multiple resources are assigned to perform tactical functions in a specified geographic area (such as on a specified floor or side of a structure), a *"Sector Division"* shall be established to provide coordination and control of tactical operations.

Designation of Sectors

When Sector Division boundaries are established on the exterior of a structure or in nonstructural incidents (such as a wildland fire), the term "side" and a number designation (1, 2, 3, 4) shall be used. In addition to establishing the Sector Division designation, specific boundaries must be defined. This is particularly important in nonstructural incidents.

When Sector Division boundaries are defined by level in a structural incident, the term "floor" and a number or descriptive designation shall be used (1, 2, 3, basement). If a Sector Division is given responsibility for the entire structure, it shall be designated as the *"Interior Sector."*

In radio communications with a Sector Division, the number designation shall follow "Sector" (Sector Side 1, Sector Floor 3). If a descriptive designation is given, it shall precede "Sector" (Interior Sector, etc.).

Sector Groups shall be designated by function (Ventilation, Fire Attack, Water Supply, etc.). In radio communications with a Sector Group, the function shall serve as the designation.

9. STAGING

When the IC has not defined an assignment for on-scene or responding resources, Staging shall be established.

Level I Staging: Units arriving after the initial attack should report to their preplanned locations and if no orders are received, it is recommended they take a hold position in the vicinity of the incident and await assignment by the IC.

Level II Staging: When an incident is escalating or has not yet been stabilized, sufficient resources to meet potential incident development should be available in Staging until the incident has been stabilized. The IC or Operations shall establish Staging by defining its location and communicating this information to the County-Wide Communications Center. The Dispatcher shall inform all responding resources of the location of Staging.

If responsibility is not specifically assigned, the Officer of the first company, to arrive in Staging shall assume the function of Staging Area Manager.

Resources in Staging shall remain integrity (remain with their company) and be available for immediate assignment and deployment.

All firefighters responding to the scene in their private vehicles shall report to their Fire Company Officer for assignment.

The Staging Area Manager shall keep the IC or Operations advised of resource availability in Staging whenever resource status changes.

The IC or Operations shall request on-scene resources through the Staging Area Manager and shall specify where and to whom those resources shall report.

In radio communications "Staging" will be the call sign designation and if incident has been named than "Staging" shall precede the incident name. Example: Willow Street Staging.

10. ICS ORGANIZATION FOR LARGER INCIDENTS

ICS organizational structure should be based on the management needs of the incident and should be developed on a proactive basis. Incident resource and management needs must be projected adequately ahead to allow for the reflex time of responding resources. (See ICS Charts.)

The IC and other supervisory personnel should anticipate span-of-control problems. Subordinate management positions should be staffed to maintain an acceptable span of control and workload. This may necessitate requesting additional command officers to fill these overhead positions. Example: Establish Branches to supervise Sectors who in turn supervise resources and task forces.

Whenever Operations, Planning, Logistics, or Finance functional responsibilities become a significant workload for the IC, they should be staffed with an Officer. Checklists for these functions plus the Safety Officer, Liaison Officer, and the Public Information Officer are located on the Fire Coordinator's Vehicle, Manheim Township Command Bus, or as requested by any Lancaster County Fire Chief.

11. TRANSFER OF COMMAND

Command may be transferred from the initial IC (often a Company Officer) to a later arriving or Senior Command Officer. Transfer of command shall take place on a face-to-face basis whenever possible to facilitate effective communication and feedback. If face-to-face communication is not possible, transfer of command by radio may be conducted.

If command has been established by a Firefighter, command shall be transferred to the first arriving Officer. Command shall be transferred to the first arriving Chief Officer at that Officer's discretion (the Chief Officer may choose to allow the Company Officer to continue as IC). Transfer of command to higher ranking command Chief Officers is also discretionary. When a Chief Officer allows a lower ranking Officer to retain command, this does not remove the responsibility for the incident from the higher ranking individual.

Transfer of command shall include communication of the following information:

1. The status of current situation.

2. Resources committed to the incident and responding, as well as the present incident organizational structure.

3. Assessment of the current effect of tactical operations.

Following transfer of command the IC may return the previous IC to his or her Company (if a Chief Officer) or specify assignment to a subordinate management position within the ICS organizational structure.

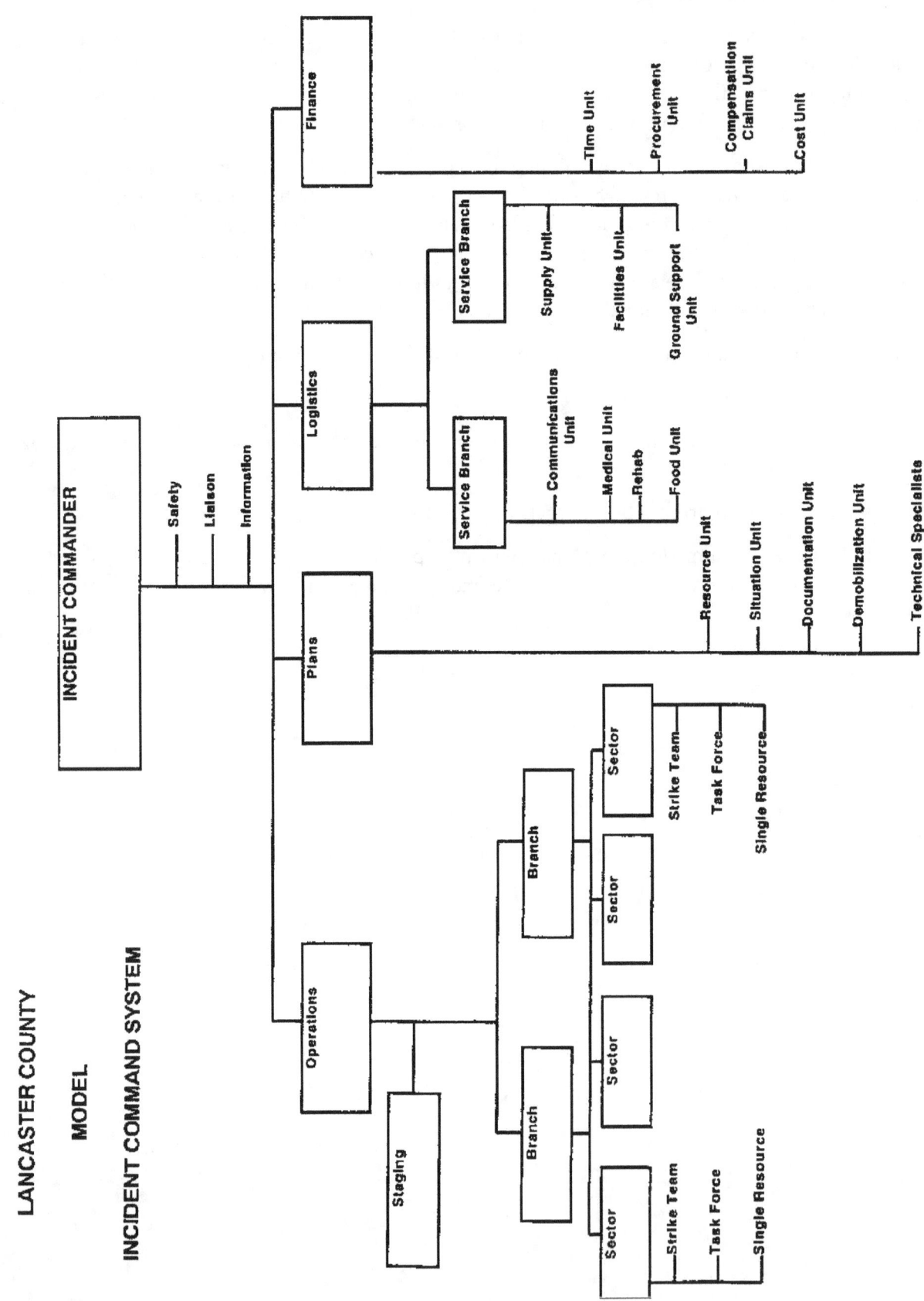

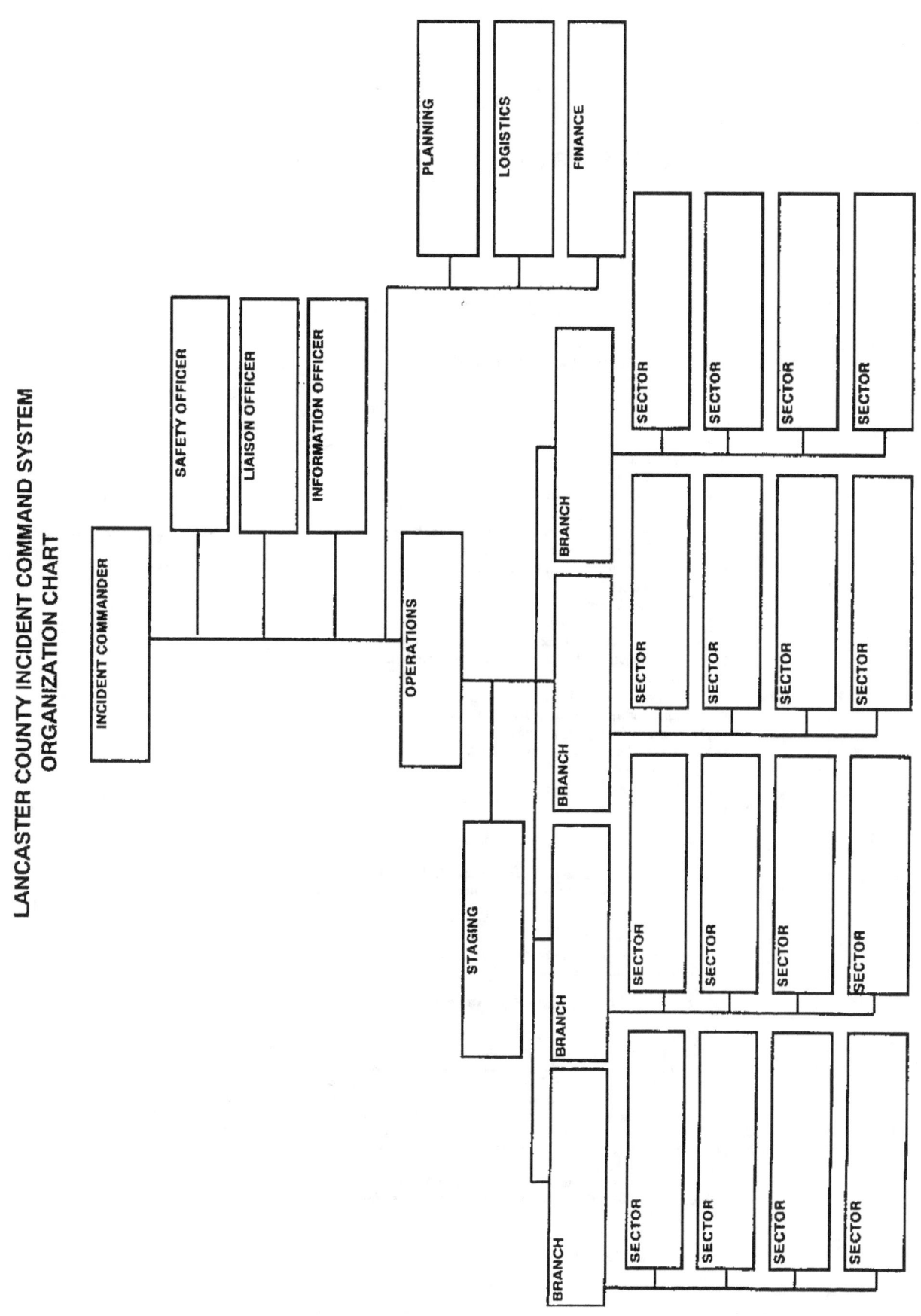

APPENDIX B

State Panic And Fire Code

Ch. 49   ADMINISTRATION—BUILDINGS

CHAPTER 49. ADMINISTRATION—BUILDINGS

Sec.
49.1. Definitions.
49.2. Jurisdiction and effective dates.
49.3. Submission of plans.
49.4. Professional registration requirements.
49.5. Certification of plans.
49.6. Appeal of Department action as to final plans.
49.7. Legal effect of approval of plans.
49.8. Fees for building-plan examinations.
49.9. Field inspection.
49.10. Applicability of general rules.
49.11. Service of orders, notices and duty of owner to post address.
49.12. Notice of violation and order to correct.
49.13. Determination of compliance or noncompliance.
49.14. Answer to order to show cause.
49.15. Appeals to the Board.
49.16. Enforcement of final order after Board action.
49.17. Cases involving danger of imminent harm.
49.18. Appeal from orders of the Board.
49.19. Advisory Board.
49.20. Information concerning protection from fire.
49.21. Reimbursement of municipalities for installation of teletypewriters.

Compliments of: Senator NOAH W. WENGER 36th Senatorial District

Authority

The provisions of this Chapter 49 issued under act of April 27, 1927 (P. L. 465, No. 299) (35 P. S. §§ 1221—1235), unless otherwise noted.

Source

The provisions of this Chapter 49 adopted May 18, 1984, effective May 19, 1984, 14 Pa.B. 1765, unless otherwise noted.

Cross References

This chapter cited in 6 Pa. Code § 21.30 (relating to special program and recertification standards for domiciliary care homes with four or more clients); 7 Pa. Code § 139.42 (relating to structures); 16 Pa. Code § 29.35 (relating to applicability of Fire and Panic Act); 22 Pa. Code § 31.43 (relating to buildings and equipment); 25 Pa. Code § 177.17 (relating to safety—fire prevention and egress); 34 Pa. Code § 11.85 (relating to applicable provisions of other regulations); 34 Pa. Code § 47.125 (relating to stairs); 34 Pa. Code § 50.21 (relating to definitions); 34 Pa. Code § 50.24 (relating to exit doors and exit access doors); 34 Pa. Code § 50.53 (relating to general fire alarm requirements); 34 Pa. Code § 50.92 (relating to historic building); 34 Pa. Code § 60.8 (relating to field inspection); 34 Pa. Code § 60.33 (relating to buildings); 55 Pa. Code § 5200.47 (relating to other applicable regulations); and 55 Pa. Code § 5210.56 (relating to other applicable regulations).

Appendix B (continued)

Ch. 51   A-1 ASSEMBLY

CHAPTER 51. A-1 ASSEMBLY
OCCUPANCY GROUP

Sec.
- 51.1. Division A-1 occupancies.
- 51.2. Separation and mixed occupancy.
- 51.3. Hazards.

CONSTRUCTION TYPE—MAXIMUM STORY HEIGHT

- 51.11. Story height.

MEANS OF EGRESS

- 51.21. Minimum exits.
- 51.22. Travel distance to exits.
- 51.23. Means of egress capacity.
- 51.24. Exit doors.
- 51.25. Stair towers.
- 51.26. Intercommunicating stairways.
- 51.27. Ramps.
- 51.28. Horizontal exits.
- 51.29. Escalators.

VERTICAL OPENINGS

- 51.31. Vertical openings.

INTERIOR FINISH

- 51.41. Interior finishes.

MANUAL AND AUTOMATIC FIRE ALARM SYSTEMS

- 51.51. Not required.

EMERGENCY LIGHTING SYSTEMS

- 51.61. General requirements.

EXTINGUISHERS AND SPRINKLER SYSTEMS

- 51.71. Fire extinguishers.
- 51.72. Automatic sprinkler systems.

SPECIAL CONDITIONS

- 51.81. Hazardous areas.

(188647) No. 238 Sep. 94

Appendix B (continued)

34 § 51.1 DEPT. OF LABOR & INDUSTRY Pt. I

Authority

The provisions of this Chapter 51 issued under act of April 27, 1927 (P. L. 465, No. 299) (35 P. S. §§ 1221—1235), unless otherwise noted.

Source

The provisions of this Chapter 51 adopted May 18, 1984, effective May 19, 1984, 14 Pa.B. 1765, unless otherwise noted.

Cross References

This chapter cited in 7 Pa. Code § 139.42 (relating to structures); 16 Pa. Code § 29.35 (relating to applicability of Fire and Panic Act); 22 Pa. Code § 31.43 (relating to buildings and equipment); 25 Pa. Code § 177.17 (relating to safety—fire prevention and egress); 34 Pa. Code § 11.85 (relating to applicable provisions of other regulations); 34 Pa. Code § 47.125 (relating to stairs); 34 Pa. Code § 49.1 (relating to definitions); 34 Pa. Code § 49.2 (relating to jurisdiction and effective dates); 34 Pa. Code § 49.3 (relating to submission of plans); 34 Pa. Code § 49.5 (relating to certification of plans); 34 Pa. Code § 49.8 (relating to fees for inspection); 34 Pa. Code § 49.9 (relating to field inspection); 34 Pa. Code § 49.12 (relating to notice of violation and order to correct); 34 Pa. Code § 50.1 (relating to occupancy groups); 34 Pa. Code § 50.21 (relating to definitions); 34 Pa. Code § 50.24 (relating to exit doors and exit access doors); 34 Pa. Code § 50.92 (relating to historic building); 34 Pa. Code § 60.8 (relating to field inspection); 34 Pa. Code § 60.33 (relating to buildings); 55 Pa. Code § 5200.47 (relating to other applicable regulations); and 55 Pa. Code § 5210.56 (relating to other applicable regulations).

OCCUPANCY GROUP

§ 51.1. Division A-1 occupancies.

Buildings primarily used or designed for the purpose of assembly of 501 or more persons for amusement, entertainment, worship, transportation, recreation, sports, military drilling, dining or similar purposes shall be classified as Division A-1 occupancies.

§ 51.2. Separation and mixed occupancy.

When an A-1 occupancy shares a structure with one or more occupancies, the structure shall be governed in one of the following manners:

(1) *Separation.* When each occupancy is separated from all other occupancies by fire walls, each portion thus separated shall be considered a separate building, and limitations for separate buildings shall govern.

(2) *Mixed occupancy.* Buildings with multiple occupancies which are not separated shall be considered mixed occupancies and shall be governed by the most restrictive limitations of the various occupancies.

§ 51.3. Hazards.

D-H, hazardous occupancies, may not be permitted in the same structure housing an A-1 occupancy.

Appendix B (continued)

Ch. 51   A-1 ASSEMBLY   34 § 51.11

CONSTRUCTION TYPE — MAXIMUM STORY HEIGHT

§51.11. Story height.

(a) A-1 occupancies shall be governed by the story height limitations in the following table:

Construction Type	Maximum Story Height
Fireresistive	No Limit
Noncombustible	1 Story
Protected Heavy Timber	1 Story
Ordinary	*Not Permitted
Wood Frame	*Not Permitted

(b) *A single story shall be permitted in ordinary or wood frame construction when the building is 1-hour rated construction or totally protected by an automatic sprinkler system.

(c) An additional story shall be permitted in noncombustible and protected heavy timber construction when the building is 1-hour rated construction or totally protected by an automatic sprinkler system. The maximum story height may be increased by two stories.

(d) Noncombustible, protected heavy timber, ordinary and wood frame construction types shall be considered 1-hour rated construction when bearing walls, columns, beams, other supporting members, ceiling/roof assemblies, floor/ceiling assemblies, stairways, and other openings through floors are of 1-hour rated construction.

(e) Basement areas shall be 1-hour rated construction up to and including the floor/ceiling assembly between the basement and the first floor.

MEANS OF EGRESS

§51.21. Minimum exits.

(a) There shall be a minimum of three exits reasonably remote from each other. No more than 50% of the required exit capacity shall exit into the same unseparated corridor network.

(b) Basements used only for storage or mechanical equipment without any permanent occupancy shall have a minimum of two exit access paths except that the Department may permit a single exit access for basements of less than 1,000 square feet which are used only for storage or mechanical equipment without any permanent occupancy.

§51.22. Travel distance to exits.

(a) Exits shall be so arranged that the total length of travel from any point to reach an exit will not exceed 150 feet. Exits shall be so arranged that one exit

Appendix B (continued)

34 § 51.23   INDUSTRIAL BOARD   Pt. I

is not more than 200 feet from another exit. Dead ends and occupancy areas with a single path of egress travel should be eliminated where possible but in no case shall any of these occupancy areas be more than 75 feet from an exit.

(b)   Travel distance may be increased to the following in buildings totally protected by an automatic sprinkler system:

(1)   200 feet from any point to an exit.

(2)   300 feet between exits.

(3)   100 feet for dead ends and areas with a single path of egress.

§51.23. Means of egress capacity.

Units of width shall comply with § 50.23 (relating to means of egress capacity).

§51.24. Exit doors.

Exit discharge doors leading to the outside shall comply with § 50.24 (relating to exit doors and exit access doors).

§51.25. Stair towers.

Stair towers shall comply with § 50.25 (relating to stair towers). Only Class A stairs may be used as a required means of egress.

§51.26. Intercommunicating stairways.

Intercommunicating stairways may be used to communicate from story to story; however, they shall not be counted as exits.

§51.27. Ramps.

Ramps shall comply with § 50.27 (relating to ramps).

§51.28. Horizontal exits.

Horizontal exits shall comply with § 50.28 (relating to horizontal exits).

§51.29. Escalators.

Escalators shall be permitted for communication from one story to another but shall not be counted as exits.

Appendix B (continued)

Ch. 51   A-1 ASSEMBLY   34 § 51.31

VERTICAL OPENINGS

§51.31. Vertical openings.

Vertical openings shall comply with §§ 50.31 — 50.34 (relating to vertical openings).

INTERIOR FINISH

§51.41. Interior finishes.

(a) Interior finishes shall be Class A for exits and exit corridors except that carpet covering floors may be Class A or B. Other interior finishes shall be Class A or Class B.

(b) Interior finish requirements may be reduced by one class in buildings totally protected by an automatic sprinkler system.

MANUAL AND AUTOMATIC FIRE ALARM SYSTEMS

§51.51. Not required.

Manual and automatic fire alarm systems are not required for A-1 occupancy buildings.

EMERGENCY LIGHTING SYSTEMS

§51.61. General requirements.

Emergency lighting shall be provided as follows:

(1) Exitways, corridors, stairways, passageways, halls, landings of stairs, exit doors, including angles and intersections, and other means of egress.

(2) Rooms used for assembly purposes in excess of 750 square feet.

(3) To illuminate exit or directional exit signs.

(4) Rooms in which emergency lighting equipment is located.

(5) Exterior light over required exit discharge.

EXTINGUISHERS AND SPRINKLER SYSTEMS

§51.71. Fire extinguishers.

(a) A minimum of one fire extinguisher with a minimum 2-A rating shall be provided for each 5,000 square feet or fraction thereof, but there shall be no less than one fire extinguisher per floor including basement. Fire extinguishers

Appendix B (continued)

34 § 51.72 INDUSTRIAL BOARD Pt. I

shall be located so that it shall not be necessary to travel more than 100 feet in any direction to reach the nearest unit.

(b) A fire extinguisher with a minimum 10-B rating shall be provided in each kitchen.

§51.72. Automatic sprinkler systems.

(a) Automatic sprinkler protection is required for A-1 occupancies in all storage rooms over 100 square feet, maintenance rooms and boiler and heater rooms except boiler and heater rooms with equipment totally operated by electricity.

(b) Buildings or structures meeting the definition of high rise buildings in § 49.1 (relating to definitions) shall be equipped with an automatic sprinkler system.

SPECIAL CONDITIONS

§51.81. Hazardous areas.

(a) Boiler or furnace rooms, repair or maintenance rooms, trash rooms, and rooms or spaces used for storage of combustible materials in quantities deemed hazardous by the Department shall be separated from other areas of the building by 1-hour partitions, floors, and ceilings. Openings shall be protected by C label door assemblies. Combustion and ventilation air for boiler, incinerator or heater rooms shall be taken directly from and discharged directly to the outside air.

(b) American Gas Association approved gas fire forced air furnaces and space heaters; U. L. approved electrical resistive coil heating furnaces and U. L. approved oil fired forced air furnaces and space heaters need not be enclosed.

(c) Aisles in auditoriums shall be provided with general illumination of not less than 0.1 foot candles at the front row of seats and not less than 0.2 foot candles at the last row of seats, and the illumination shall be maintained throughout the showing of motion pictures or other projections.

[Next page is 52-1.]

APPENDIX C

Emergency Services Operating Costs

Sight and Sound Theater – January 28, 1997

FIRE SERVICE

Pumpers: 8 for 7 hr. & 5 for an additional 10 hr. for overhaul and reburns.

$$1 \text{ hr.} \times \$200 \times 8 = \$1{,}600$$
$$6 \text{ hr.} \times \$150 \times 8 = \$7{,}200$$
$$10 \text{ hr.} \times \$150 \times 5 = \$7{,}500$$
$$\text{Sub Total} = \$16{,}300$$

Ladder Trucks: 5 for 7 hr. & 1 for an additional 10 hr. for overhaul and reburns.

$$7 \text{ hr.} \times \$200 \times 5 = \$7{,}000$$
$$10 \text{ hr} \times \$200 \times 1 = \$2{,}000$$
$$\text{Sub Total} = \$9{,}000$$

Water Tankers: 15 for 7 hr. & 3 for an additional 10 hr. for overhaul and reburns.

$$1 \text{ hr.} \times \$200 \times 15 = \$3{,}000$$
$$6 \text{ hr.} \times \$150 \times 15 = \$13{,}500$$
$$10 \text{ hr.} \times \$150 \times 3 = \$4{,}500$$
$$\text{Sub Total} = \$21{,}000$$

Squads: 5 for 7 hr. & 1 for an additional 10 hr. for overhaul and reburns.

$$7 \text{ hr.} \times \$150 \times 5 = \$5{,}250$$
$$10 \text{ hr.} \times \$150 \times 1 = \$1{,}500$$
$$\text{Sub Total} = \$6{,}750$$

Rescue: 2 for 7 hr.

$$7 \text{ hr.} \times \$200 \times 2 = \$2{,}800$$
$$\text{Sub Total} = \$2{,}800$$

Hazmat: 1 for 4 hr.

$$4 \text{ hr.} \times \$200 \times 1 = \$800$$
$$\text{Sub Total} = \$800$$

conitinued on next page

(Fire Service continued from previous page)

Fire Police: 5 Traffic Control Points for 14 hr.

$$5 \times 12 = \underline{\$60}$$
$$\text{Sub Total} = \$60$$

Firefighters' lost wages: 200 for 7 hr. & 50 for an additional 10 hr. for overhaul and reburns.

$$7 \text{ hr.} \times \$8 \text{ per hour} \times 200 = \$11,200$$
$$10 \text{ hr.} \times \$8 \text{ per hour} \times 50 = \underline{\$4,000}$$
$$\text{Sub Total} = \$15,200$$

Total Fire Service = $71,910

EMERGENCY MEDICAL

Ambulances: 6 for 10 hr.

$$10 \times \$150 \times 6 = \underline{\$9,000}$$
$$\text{Sub Total} = \$9,000$$

EMS wages: 20 personnel for 10 hr.

$$10 \times \$8 \text{ per hour} \times 20 = \underline{\$1,600}$$
$$\text{Sub Total} = \$1,600$$

Total EMS = $10,600

AMERICAN RED CROSS

Support/Canteen Unit: 1 for 10 hr.

$$10 \times \$150 \times 1 = \underline{\$1,500}$$
$$\text{Sub Total} = \$1,500$$

Personnel: 2 for 10 hr.

$$10 \text{ hr.} \times \$8 \text{ per hour} \times 2 = \underline{\$160}$$
$$\text{Sub Total} = \$160$$

Food & Drinks: 1 per 220 personnel

$$1 \text{ meal} \times \$10 \times 220 = \underline{\$2,200}$$
$$\text{Sub Total} = \$2,200$$

Total ARC = $3,860

TOTAL COSTS: $86,370

Note: The total direct costs of the operation are based on information available to the Fire Service Coordinator, and in some figures are based on averages. Actual total costs may be higher or lower than indicated.

APPENDIX D

Lancaster Fire Chief Association
Standard Operating Guidelines

SAFETY PROCEDURES LANCASTER COUNTY, PA.

STANDARD OPERATING GUIDELINES Revised 12/17/94

PERSONNEL ACCOUNTABILITY Lancaster County Chiefs Association

This procedure describes the steps and responsibilities necessary for maintaining accountability for all personnel operating at emergency incidents.

Accountability is directly related to supervision. It is the responsibility of all Company Officers, Sector Officers, and Command to maintain a level of supervision that accounts for the location and function of all personnel at every incident.

Company Officers will keep their crew intact, maintain a constant awareness of crew members' welfare, and maintain a means to communicate with Command.

It will be the responsibility of individual firefighters and other personnel at the incident site to keep their supervisors informed of their activities and whereabouts. Freelancing of activities will not be permitted and can lead to injury and death of the firefighter and others. Freelancing is defined as individual activities carried out independently of direct or indirect orders from the Incident Command or Sector Officers.

EMERGENCY INCIDENT ACCOUNTABILITY

All crews and crew members will remain intact and under supervision by a Company or Command Officer.

All crews or other personnel assigned to incident duties mast have a radio for communicating with Command.

All crews working within a hazardous area shall utilize at least a two-person buddy system.

All PASS devices must be turned to the "on" position when operating on the fireground. Should a firefighter become lost or injured, he/she shall immediately turn the PASS device to the "distress" position to sound the alarm.

EMERGENCY EVACUATION

When evacuating a situation under "emergency" conditions, the Company Officer will immediately collect all his/her crew members, or other personnel assigned, and retreat as quickly as possible. Once in a safe area, the Company Officer will conduct an immediate roll call of all personnel assigned to his/her supervision. The Company Officer will report the roll call results (i.e., "accounted for" or "firefighter missing") to the appropriate Sector Officer or Command.

The Incident Commander shall designate a specific frequency/channel for communications between the Command Post and Sectors. All members operating on this channel/frequency shall immediately cease communications upon the transmission of "MAYDAY" or "URGENT," or any other language indicating an emergency condition. Immediately following the 'emergency" evacuation order, Command will confirm with each Sector that all Sectors copied the evacuation order.

Sector Officers will contact each company within the hazard area to confirm that the company copied the order to evacuate are indeed withdrawing.

Upon reaching a safe area, Sector Officers will contact all Company Officers to confirm the roll call results. Each Sector Officer will advise Command of roll call results.

Command will be responsible for confirming that all personnel that have been assigned to incident duties are accounted for.

The Incident Commander will record the results of the "Roll Call" and status of any missing firefighters, and will immediately transmit an additional alarm for manpower upon receipt of "Firefighter Missing," and shall notify County Communications to mark the time of such notification, and any further changes in the update of the missing firefighter (such as FIREFIGHTER LOCATED, ADDITIONAL FIREFIGHTERS MISSING, FIREFIGHTER REMOVED, NOTIFY CORONER, ETC.).

LANCASTER COUNTY FIREFIGHTER ACCOUNTABILITY SYSTEM

Firefighter ID Tag

Each firefighter shall be issued one accountability ID tag. This tag shall consist of a laminated identification tag attached to the firefighter's helmet or coat by a metal snap.

Unit Collector Ring

Each piece of apparatus shall carry one Unit Collector Ring which shall consist of a large metal ring with a 4" yellow plastic vehicle ID tag attached to it (i.e.: E 2-5-1).

Accountability Binder

An Accountability Binder shall be used by the Incident Commander or his designated Accountability Officer to display the ID Tags in a well-organized manner.

LEVEL 1 ACCOUNTABILITY

The purpose of Level 1 Accountability is so establish a primary method of accounting for all personnel on the fireground.

A Level 1 Accountability System shall be implemented on every incident, regardless of its size or nature.

LEVEL 1 ACCOUNTABILITY SHALL CONSIST OF THE FOLLOWING:

1. Upon boarding the apparatus, all personnel shall remove his/her laminated ID tag and place it on the apparatus in the position determined by his/her Fire Company/Department.

2. Level 1 Accountability shall remain in effect until the apparatus is released and returning to station or until the Incident Commander determines that Level 2 Accountability is necessary.

LEVEL 2 ACCOUNTABILITY

Level 2 Accountability is to establish a comprehensive method of accounting for all personnel on the fireground. Level 2 Accountability organizes personnel in "Companies" and discourages "freelancing."

LEVEL 2 ACCOUNTABILITY SHALL CONSIST OF THE FOLLOWING:

The Incident Commander initiates Level 2 Accountability when he/she determines that hazardous conditions exist at the incident scene or that hazardous conditions are imminent. Hazardous conditions shall be defined as, but not limited to: severe fire conditions; possible building collapse; hazardous materials; confined space emergencies; water rescue explosion; or anytime SCBA or special protective equipment are used.

Upon activation of Level 2 Accountability the driver operator of each apparatus will collect his/her crewmembers' ID Tags and place them on the Unit Collector Ring. To insure standardization the Unit Collector Ring shall then be placed on the apparatus turn signal handle for collection by the accountability officer.

The Incident Commander shall assign an "Accountability Officer" to gather all collector rings with ID Tags from each unit at the scene and in staging. The collector rings with ID Tags shall be placed in the Accountability Binder. Next to each clip, the Accountability Officer shall write the Unit ID) Number, Unit Location, and Unit Function.

Personnel reporting to the scene in private vehicles shall report to the Command Post for an assignment and shall be "checked in" to the accountability system. This process will discourage "freelancing," improve safety, and aid in tracking and identification of all personnel operating at the scene and in staging

Firefighter accountability is an ongoing process. Sector Officers shall keep the Accountability Officer appraised of any and all personnel changes as well as changes in unit status or assignment. The Accountability Officer shall brief the Incident Commander periodically on personnel status.

Spare, additional, or blank tags may be obtained from the Fire Service Coordinator.